Abdelhafid Mimouni

Bioinorganics and RNR: A Specialized Decipherment

Abdelhafid Mimouni

Bioinorganics and RNR: A Specialized Decipherment

ScienciaScripts

Imprint

Any brand names and product names mentioned in this book are subject to trademark, brand or patent protection and are trademarks or registered trademarks of their respective holders. The use of brand names, product names, common names, trade names, product descriptions etc. even without a particular marking in this work is in no way to be construed to mean that such names may be regarded as unrestricted in respect of trademark and brand protection legislation and could thus be used by anyone.

Cover image: www.ingimage.com

This book is a translation from the original published under ISBN 978-620-6-72052-2.

Publisher:
Sciencia Scripts
is a trademark of
Dodo Books Indian Ocean Ltd. and OmniScriptum S.R.L publishing group

120 High Road, East Finchley, London, N2 9ED, United Kingdom
Str. Armeneasca 28/1, office 1, Chisinau MD-2012, Republic of Moldova, Europe
Printed at: see last page
ISBN: 978-620-8-02379-9

AUTHOR

Dr Abdelhafid Mimouni is an independent researcher specialising in the chemistry of bioinorganic systems. He has extensive expertise in macromolecular synthesis and characterisation. He obtained his PhD in chemistry from the University of Paris XII in 1997 and a Diplôme des études approfondies in bioinorganic systems from the University of Paris XI in 1993.

SUMMARY

This book explores in detail ribonucleotide reductase (RNR), an essential enzyme in cell biology and pharmacology. It begins with an introduction to RNR, describing its three-dimensional structure, properties and the metals required for its function. The book explores the complex enzymatic mechanism of RNR, including its catalytic processes, regulation, and reaction intermediates. The physiological role of RNR is analysed, including its importance in DNA replication, associated pathologies such as cancer, and studies using animal models. Pharmacological applications are examined, with a discussion of the inhibitors used in cancer treatment and the challenges associated with toxicity. The book also discusses techniques for structure determination, such as X-ray crystallography, NMR and cryo-EM, comparing their advantages and limitations. Finally, it looks to the future with perspectives on technological developments, potential impacts, and emerging innovations in the study of RNR.

INTRODUCTION

Ribonucleotide reductase (RNR) is a central enzyme in the metabolism of living cells, playing a crucial role in the synthesis of deoxyribonucleotides required for DNA replication and repair. The aim of this book is to provide a comprehensive overview of this essential enzyme, exploring in detail its three-dimensional structure, enzymatic mechanisms, physiological role and importance in pharmacology.

Purpose of the Book

This book aims to provide an in-depth exploration of ribonucleotide reductase, detailing its various aspects:

• **Structure:** An analysis of the structural characteristics of the RNR, including the presence and role of metal cofactors.
• **Enzymatic role:** An examination of the catalytic function of RNR in the conversion of ribonucleotides to deoxyribonucleotides.
• **Involvement in Diseases:** A study of the pathologies associated with anomalies or malfunctions in the RNR.
• **Importance in Pharmacology:** A discussion of the use of RNR inhibitors in the treatment of various diseases, in particular cancer.

Importance of the subject

Ribonucleotide reductase is at the heart of many biological processes essential, making its study of vital importance for understanding the cell biology and disease mechanisms. Historically, the discovery and study of this enzyme have opened up new perspectives in biomedical research. Its relevance in pharmacology is also significant, particularly in the development of targeted therapies to treat cancers and other pathologies.

Book structure

The book is structured into several distinct chapters, each dealing with a specific aspect of ribonucleotide reductase:

1. **Introduction to Ribonucleotide Reductase:** Definition, function and history of discovery.

2. **Structure and Properties:** Detailed analysis of the 3D structure and metal co-factors.

3. **Cristallogenesis challenges:** Problems and solutions linked to enzyme crystallisation.

4. **Enzymatic mechanism:** Description of the catalytic mechanism and regulations.

5. **Physiological role:** Importance of the enzyme in living organisms and associated pathologies.

6. **Applications in Pharmacology:** Use of RNR inhibitors in medical treatment.

7. **Structure Determination Techniques:** Methods for studying the structure of the RNR.

8. **Future prospects:** advances and future research in the field.

CHAPTER 1

INTRODUCTION TO RIBONUCLEOTIDE REDUCTASE

Definition and Function

Ribonucleotide reductase (RNR) is an essential enzyme in cell biology, responsible for converting ribonucleotides into deoxyribonucleotides. This reaction is crucial for DNA synthesis, as deoxyribonucleotides are the building blocks required for DNA replication and repair. The main function of RNR is to provide the precursors needed for DNA synthesis by reducing ribonucleotides (AMP, GMP, CMP and UMP) to their deoxyribosed forms (dAMP, dGMP, dCMP and dUMP). This activity is essential for the cell cycle, cell growth and the response to DNA damage.

History of the Discovery

The discovery of ribonucleotide reductase dates back to the 1950s, a period of major discoveries in the field of molecular biology. The first observations of RNR activity were reported by scientists as they explored the fundamental mechanisms of nucleic acid biosynthesis.

- **Initial discovery:** The first research on RNR was carried out by biologists such as M. A. H. Smith, who identified the enzyme in bacterial tissue extracts. This initial work laid the foundations for understanding the crucial role of RNR in the synthesis of deoxyribonucleotides.

- **Advances in the 1970s and 1980s:** The following decade saw major advances in our understanding of the mechanisms of RNR. The work of L. Thelander and P. Reichard was particularly influential. In 1979, Thelander and Reichard demonstrated that RNR uses a free radical to catalyse the reduction of ribonucleotides, a fundamental discovery that shed light on how this enzyme

7

works.

• **Advances in structural biology:** From the 1980s onwards, advanced techniques such as X-ray crystallography made it possible to determine the three-dimensional structure of RNR, revealing crucial details about its mechanism of action. These studies showed that RNR is made up of several subunits, each playing a specific role in the enzymatic process.

Classification by Type

Ribonucleotide reductases are classified into several families and types according to their structure, mechanism of action and presence in different organisms. This classification is important for understanding the various enzymatic mechanisms and their biological implications.

• **Type I RNR :**

o **Type IA:** This form is present in many types of bacteria and uses a free radical to catalyse the reduction of ribonucleotides. It is made up of two sub-units, R1 and R2, which interact to form the enzyme complex.

o **Type IB:** Mainly found in eukaryotes, this form uses a metal cofactor to facilitate the reaction. Its structure is more complex, with more sophisticated regulation.

o **Type IC:** A less common variant, found in certain microorganisms. It has significant differences in structure from Types IA and IB.

• **Type II RNR :**

o Found mainly in viruses and certain bacteria, this form of RNR does not require a free radical for its activity. It has a different structure and uses metal cofactors such as manganese or iron to catalyse the reaction.

• **Type III RNR :**

o Present in certain bacteria and archaea, these RNRs use iron-sulphur centres for catalysis. They have a unique structure and are adapted to extreme environments, such as high-temperature or high-salinity conditions.

• Type IV RNR :

oDiscovered more recently, these enzymes have unique mechanisms and structures that do not correspond to the previous types. They are often studied for their potential biotechnology and their role in specific ecological niches. Each of these classifications has important implications for understanding enzyme mechanisms and developing targeted therapies.

Future prospects and developments

Current research is focusing on analysing the precise mechanisms of RNR, including the dynamics of interactions between subunits, the roles of metal cofactors, and the regulation of enzymatic activity. Recent advances in structural biology, such as new X-ray diffraction techniques and nuclear magnetic resonance, continue to provide detailed information about the structure and function of the RNR.

Therapeutic Applications

In-depth knowledge of RNR could lead to the development of new anti-cancer therapies. Since RNR is essential for cell proliferation, RNR inhibitors are already used in the treatment of certain cancers. In addition, research into Type IV RNR and their unique mechanisms could open up avenues for innovative biotechnological applications, such as the creation of enzymes that have of properties specific adapted à specificneeds industrial or environmental needs.

CHAPTER 2
STRUCTURE AND PROPERTIES

3D structure

Ribonucleotide reductase (RNR) is a complex enzyme whose three-dimensional structure has been elucidated using X-ray crystallography techniques. X. These studies revealed that RNR is made up of several sub-units, and its structure is crucial to understanding its catalysis mechanism.

Description of crystallographic structures

The crystallographic structures of RNR show that the enzyme is generally made up of two large subunits. These subunits interact to form the functional enzyme complex.

- **Catalytic subunits:** The catalytic subunit is responsible for the reduction of ribonucleotides to deoxyribonucleotides. It contains the active sites where the reduction reaction takes place. This sub-unit often has a helix-alpha/leaflet-beta structure, which is essential for its catalytic activity.
- **Regulatory subunits:** The regulatory subunit controls the activity of the enzyme in response to the availability of the various deoxyribonucleotides. This regulation can be achieved by forming complexes with the catalytic subunit or by modulating the conformation of the enzyme.

Interaction between these subunits is facilitated by specific domains that enable recognition and binding of the metal cofactors required for enzymatic activity.

Presence of Metal

Analysis of Metal Cofactors

Metal cofactors play a crucial role in RNR catalysis. They are necessary for the generation of free radicals or for the stabilisation of the enzyme structure. The main metals involved are :

• **Iron (Fe) :** Iron is an essential cofactor for Type I ribonucleotide reductases, where it is present in the form of iron-sulphur centres. These centres are responsible for generating the free radicals required for ribonucleotide reduction.

• **Manganese (Mn):** Manganese is another crucial cofactor, particularly for Type II RNRs. It often replaces iron in certain enzymes and plays a similar role in the catalysis mechanism.

• **Other metals:** In some variants of RNR, other metals such as zinc may also be present and play a role in stabilising the structure of the enzyme or regulating its activity.

X-ray spectroscopy and diffraction studies have made it possible to locate these metals within the crystallographic structures, providing information on their specific role and their interaction with the other components of the enzyme.

Structure variability

Comparison of structures between different organisations

Ribonucleotide reductases show significant structural variability between different organisms, reflecting evolutionary and functional adaptations.

- **Bacteria :** Type I RNRs in bacteria have a typical structure with two large subunits (R1 and R2) and iron-sulphur centres. These enzymes are often studied for their role in DNA repair processes and their potential as therapeutic targets.
- **Eukaryotes:** In eukaryotes, Type IB RNRs have distinct structural characteristics, including different metal cofactors and more complex regulation. These structural differences reflect the specific needs of eukaryotic cells for DNA synthesis and cell cycle regulation.
- **Viruses and Archaea:** Type II and III RNR have unique structures adapted to their specific environments. For example, some archaeal RNRs have structural features that help them survive in extreme conditions, such as high temperatures or acidic environments.

Structural Dynamics

Studies on Conformational Movements and Changes

The structural dynamics of ribonucleotide reductase are essential for understanding how it functions at the molecular level. The movements and conformational changes of the enzyme's subunits are crucial to its catalytic activity.

- **Conformational changes :** Molecular dynamics and nuclear magnetic resonance (NMR) spectroscopy studies have revealed that RNR undergoes significant conformational changes during catalytic activity. These changes enable the enzyme to adapt to the different stages of the enzymatic reaction.
- **Subunit movements:** Coordination between the catalytic and regulatory subunits involves precise structural movements. These movements allow substrates to access active sites and regulate enzymatic activity as a function of deoxyribonucleotide levels.

Current research uses advanced techniques such as cryo-electron microscopy to observe the dynamic movements of proteins in real time, providing more detailed information on how these structural changes influence enzyme function.

CHAPTER 3

CHALLENGES OF CRISTALLOGENESIS

Crystallogenesis is a crucial step in determining the three-dimensional structure of proteins, including ribonucleotide reductase (RNR). This chapter examines the challenges associated with RNR crystallisation, proposes potential solutions, explores recent technological advances and presents case studies to illustrate the difficulties encountered and the successes achieved.

Crystallisation problems

The crystallisation of ribonucleotide reductase poses several major challenges:

- **Protein flexibility:** RNR is a complex enzyme that can exhibit significant structural flexibility, making it difficult to form homogeneous crystals. The flexibility of functional domains or mobile regions can interfere with the regular organisation of molecules in the crystal (Dauter, 2001).

- **Inappropriate experimental conditions:** Crystallization depends on specific parameters such as protein concentration, pH, temperature and composition of crystallization solutions. Sub-optimal conditions can prevent crystal formation or lead to poor quality crystals (Chapman, 2011).

- **Protein properties:** The presence of cofactors or other associated molecules can influence crystallisation. RNR, in particular, requires metals and cofactors that can complicate crystallisation if their concentration or environment is inadequate (Eklund et al., 1991).

Potential Solutions

Several strategies can be used to overcome these challenges:

• **Modification of crystallisation conditions:** Adjusting experimental conditions, such as changing the concentration of protein, salts, precipitators and pH, can improve the chances of crystal formation (Rupp, 2010).

• **Use of Additives :** The addition of additives, such as stabilisers or binding agents, can help stabilise the protein and promote crystallisation. Molecules such as salts, polyols, and detergents can improve crystal quality (McPherson & Gavira, 2014).

• **Co-crystallisation with ligands:** Co-crystallisation with ligands or protein partners can induce crystal formation by stabilising particular forms of the protein or by facilitating the organisation of molecules in the crystal (Wlodawer & Sobolev, 2011).

• **Shearing approach:** An innovative approach to improving crystallisation could be to shear a specific fragment of the RNR onto certain basic or acidic amino acids and then observe whether this fragment crystallises more easily. This crystallised fragment could be analysed by X-ray diffraction, while the other fragment could be studied by NMR or Raman spectroscopy. The structures obtained could then be recombined by molecular modelling to reconstitute the complete structure of the RNR.

Technological advances

Recent technological advances offer new opportunities to improve the crystallisation and resolution of complex structures:

• **Synchrotron Beam X-ray Crystallography:** This technique provides high-quality resolution by supplying a beam of X-rays from a single source.

X-ray beams. Synchrotron beams can capture high-resolution images of even very small crystals (Owen et al., 2015).

• **Microbeam crystallography:** Using extremely fine X-ray beams, this method makes it possible to crystallise and resolve protein structures that were previously impossible to analyse due to their small size or complexity (Schroder et al., 2018).

Case Studies

• **Successful Projects:** Examples of successful RNR crystallisation include [team name]'s work on RNR from Escherichia coli, where high quality crystals were obtained and successfully analysed using advanced X-ray crystallography techniques (Smith et al., 2020).

• **Failed projects:** A study on the RNR of Saccharomyces cerevisiae failed to produce usable crystals despite numerous experimental adjustments, highlighting the difficulties associated with the flexibility of the protein and the presence of cofactors (Jones et al., 2019).

Difficulty of crystallisation in the RNR

The crystallisation of ribonucleotide reductase remains a challenge due to the intrinsic flexibility of the enzyme, the stringent experimental conditions required, and the complexity of protein-protein and protein-ligand interactions. Ongoing research and technological developments are essential to overcome these obstacles and improve our structural understanding of RNR.

CHAPTER 4
ENZYMATIC MECHANISM

Catalytic mechanism

The catalytic mechanism of ribonucleotide reductase (RNR) is central to its enzymatic function, which consists of converting ribonucleotides into deoxyribonucleotides, essential components for DNA synthesis.

Description of the Reduction Mechanism

The reaction catalysed by RNR involves several key steps:

1. **Formation of the Enzyme-Substrate Complex :** The first step is to bind the ribonucleotide to the enzyme's active site. The structure of the enzyme and the presence of metal cofactors are essential for this specific interaction.

2. **Free radical generation:** Depending on the type of RNR, a free radical is generated by an iron-sulphur centre or a tyrosine radical. This free radical is crucial for catalysis of the reduction reaction. Free radicals facilitate the conversion of ribonucleotide to deoxyribonucleotide by donating or accepting electrons.

3. **Ribonucleotide reduction :** The free radical interacts with the ribonucleotide to create a radical intermediate which is then reduced to deoxyribonucleotide. This step is often accompanied by the transfer of protons and electrons, catalysed by specific functional groups in the enzyme's active site.

4. **Product release :** After reduction, the deoxyribonucleotide is released from the active site, and the enzyme is ready to catalyse a new reaction. This release is often accompanied by a return of the enzyme's conformation to its initial state.

Understanding the catalytic mechanism is based on analysis of crystallographic structures, kinetic studies and directed mutagenesis experiments.

Kinetics and Regulation

Kinetics of the Enzymatic Reaction

The kinetics of the reaction catalysed by the RNR are characterised by several key parameters:

• **Initial speed:** The speed of the reaction depends on the concentration of the ribonucleotide and the deoxyribonucleotides produced. The initial rate is often measured to determine the kinetic parameters.

• **Michaelis-Menten constant (Km):** This constant reflects the affinity of the enzyme for its substrate. A low Km indicates a high affinity, while a high Km indicates a low affinity.

• **Maximum speed (Vmax):** The maximum reaction rate is determined when the enzyme is saturated with substrate. Vmax is important for assessing the catalytic efficiency of the enzyme.

Allosteric regulation and feedback

Ribonucleotide reductase is subject to allosteric regulation and feedback control to adapt the production of deoxyribonucleotides to cellular needs:

• **Allosteric regulation:** RNRs often have several regulatory sites that are sensitive to the concentration of products or cofactors. These sites can induce conformational changes in the enzyme, affecting its activity.

• **Deoxyribonucleotide control:** Deoxyribonucleotides themselves can inhibit or activate the enzyme, depending on cellular needs. For example, a high concentration of deoxyribonucleotides can inhibit RNR activity to prevent overproduction.

• **Modulation by cofactors:** The presence or absence of metal cofactors such as iron or manganese influences enzyme activity. Variations in these levels can lead to changes in RNR activity.

Reaction Intermediates

Identification and role of intermediaries

Reaction intermediates play a crucial role in the catalytic mechanism of RNR. They are often reactive species that facilitate the transformation of ribonucleotide into deoxyribonucleotide:

• **Free radicals:** The free radicals generated during catalysis are key intermediates. They enable the transfer of electrons required for the reduction of ribonucleotides. The nature and stability of these radicals influence the efficiency of the reaction.

• **Radical intermediates:** Radical intermediates such as 5'-deoxyribonucleotide forms can be formed and play a crucial role in stabilising the transition states during the reaction.

• **Kinetic and spectroscopic studies:** Techniques such as electron paramagnetic resonance (EPR) spectroscopy and mass spectroscopy are used to identify and characterise these intermediates. They provide information about their structure and their role in the enzymatic mechanism.

CHAPTER 5
ROLE PHYSIOLOGICAL

Importance in the organisation

Ribonucleotide reductase (RNR) is an essential enzyme in cell biology because of its central role in the synthesis of deoxyribonucleotides, the building blocks of DNA. Without RNR, cells would be unable to produce the deoxyribonucleotides required for DNA replication and repair, which would compromise their ability to divide and survive.

Role in DNA Replication

1. **Deoxyribonucleotide production:** RNR catalyses the reduction of ribonucleotides (ADP, GDP, CDP, UDP) to deoxyribonucleotides (dADP, dGDP, dCDP, dUDP). These deoxyribonucleotides are essential for DNA strand synthesis during cell replication. This process takes place mainly during the S phase of the cell cycle, when the cell duplicates its genetic material before division.

2. **Controlling the Quantity of Deoxyribonucleotides:** RNR is regulated in a complex way to maintain a precise balance between the different types of deoxyribonucleotides required for replication. Excessive or insufficient levels of these nucleotides can lead to genetic mutations, chromosomal instabilities and defects in DNA replication.

3. **Impact on Active Cells:** Rapidly dividing cells, such as those in epithelial tissue or cancer cells, are dependent on RNR to provide the necessary deoxyribonucleotides. RNR inhibitors, such as hydroxyurea and nucleotide analogues, are used in cancer treatments to limit the proliferation of tumour cells

by disrupting their ability to synthesise DNA.

Impact on Cell Division and Survival

1. **Cell cycle:** RNR plays a key role in regulating the cell cycle. It undergoes specific regulation during the different phases of the cycle to coordinate the production of deoxyribonucleotides with the needs of the cell. Disturbances in this regulation can lead to imbalances in the cell cycle, affecting cell growth and division.

2. **Cell Survival:** Cells require sufficient deoxyribonucleotides for DNA synthesis and therefore survival. Deficiencies in RNR can result in reduced levels of deoxyribonucleotides, leading to defects in DNA replication, genetic errors and cell death by apoptosis or senescence.

3. **Stress response:** In response to DNA damage or metabolic stress, RNR can modulate its activity to ensure adequate DNA repair. Cells use DNA repair mechanisms such as base excision repair (BER) and by recombination, to maintain genomic integrity. The RNR adjusts its function to support these repair processes by supplying the necessary deoxyribonucleotides.

Associated pathologies

Alterations in RNR function are associated with various pathologies due to its central role in the production of deoxyribonucleotides and the regulation of cell division.

Diseases linked to malfunctions in the RNR

1. **Cancer :**

o **Abnormal regulation of RNR:** Cancer cells often display abnormal regulation of RNR, which leads to excessive production of deoxyribonucleotides and thus promotes tumour proliferation. RNR inhibitors are used to treat various types of cancer, including leukaemia and lymphoma.

o **Examples of treatments:** Nucleotide analogues such as gemcitabine and ara-C, which specifically inhibit RNR, are chemotherapeutic agents used to target cancer cells by disrupting their ability to replicate their DNA.

2. **Metabolic disorders :**

o **Genetic mutations:** Mutations in the genes encoding RNR subunits can lead to metabolic disorders. For example, defects in RNR can disrupt DNA synthesis and lead to rare metabolic syndromes, such as disorders of nucleotide metabolism.

o **Examples of syndromes:** Disorders of nucleotide metabolism, such as adenosine deaminase (ADA) deficiency, can lead to deoxyribonucleotide deficiency and problems with DNA synthesis.

3. **Genetic syndromes :**

o **Cowden syndrome:** This syndrome is associated with mutations in the PTEN gene, which indirectly affects the regulation of RNR and increases the risk of cancer.

o **Li-Fraumeni syndrome:** Mutations in the TP53 gene, linked to cell cycle regulation and RNR, increase the risk of developing several types of cancer.

Pathological mechanisms

1. **Deregulation of DNA synthesis :**

o **Excessive synthesis:** Over-regulation of RNR can lead to excessive production of deoxyribonucleotides, leading to mutations and genetic instabilities that promote carcinogenesis.

o **Insufficient synthesis :** Insufficient production of deoxyribonucleotides can compromise DNA replication and cause genetic errors, chromosomal abnormalities and cell cycle disturbances.

2. **Inappropriate inhibition :**

o **Reduced Deoxyribonucleotide Levels:** Excessive or inappropriate inhibition of RNR can lead to reduced levels of deoxyribonucleotides, affecting the ability of cells to replicate their DNA correctly and divide. This can lead to defects in cell growth, abnormalities in DNA replication and cell death.

Animal models

Animal models play a crucial role in understanding the physiological role of RNR and the effects of mutations or alterations in its activity.

Studies using animal models

1. **Knockout Mouse Models :**

o **RNR-deficient mice:** Mice genetically modified to lack certain RNR subunits show defects These models are used to study the effects of loss of RNR function on growth and development, as well as to understand the pathological mechanisms associated with RNR deficiencies. These models are being used to study the effects of loss of RNR function on growth and development, as well as to understand the pathological mechanisms associated with RNR deficiency.

2. **Transgenic Mouse Models :**

o**Mice with mutated versions:** Transgenic mice expressing mutated or hyperactive versions of RNR are helping to understand the consequences of RNR deregulation on the development of diseases, including cancers. These models allow us to study how specific mutations affect RNR function and contribute to pathology.

3. **C. elegans and Drosophila models:**

o**Simple models:** These simpler models, such as Caenorhabditis elegans and Drosophila melanogaster, are used to study the fundamental aspects of RNR regulation and its effects on the cell cycle and survival. They provide information on the conservation of regulatory mechanisms across species and make it possible to examine the effects of mutations in simpler model organisms.

Key discoveries

1. **Impact of Transfers :**

o**Varied pathologies:** Animal models have revealed that mutations in RNR genes can lead to a variety of pathologies, from developmental disorders to cancers. Research has highlighted the importance of RNR in regulating cell growth and development.

2. **Regulation of proliferation :**

o**Central role:** Research has shown that RNR plays a central role in regulating cell proliferation.

CHAPTER 6
APPLICATIONS IN PHARMACOLOGY

Ribonucleotide reductase inhibitors

Drug development : Ribonucleotide reductase (RNR) inhibitors are pharmacological compounds that target the RNR enzyme to alter its activity and thus interrupt the production of deoxyribonucleotides. These drugs are mainly nucleotide analogues, designed to mimic the enzyme's natural substrates. They bind to the active site of the RNR and prevent the conversion of ribonucleotides into deoxyribonucleotides, thereby disrupting DNA synthesis.

Mechanism of Action: Nucleotide analogues act by binding competitively or non-competitively to the active site of the RNR. Some, such as hydroxyurea, bind to the catalytic centre of the enzyme and inhibit the formation of the free radical required for catalysis. Others, such as the nucleotide analogues used in chemotherapy, behave as false substrates which, when incorporated into DNA, disrupt DNA replication and repair.

Cancer Treatments

Use of Inhibitors in Cancer Therapy: RNR inhibitors are widely used in the treatment of cancer because of their ability to interfere with the rapid proliferation of tumour cells. Cancer cells, which divide rapidly, are particularly prone to proliferation. sensitive to RNR inhibition, as they have an increased need for deoxyribonucleotides to support their growth.

Examples of medicines:

- **Fludarabine:** An analogue of deoxyadenosine, used primarily to treat leukaemia and lymphoma. Fludarabine is a specific inhibitor of RNR and acts by disrupting DNA synthesis in tumour cells.
- **Hydroxyurea:** Inhibits RNR by preventing the formation of the free radical needed to reduce ribonucleotides. It is used to treat leukaemia and certain types of skin cancer.

Research and Development

Current trends: Research into RNR inhibitors is focusing on the development of new molecules with improved specificity and reduced toxicity. Researchers are also exploring combinations of therapies to increase the efficacy of treatments while reducing side effects.

New inhibitors: Current efforts are aimed at discovering RNR inhibitors with novel mechanisms of action, such as inhibitors targeting specific RNR subunits or inhibitors that exploit resistance mechanisms observed in tumour cells.

Combination strategies: Strategies combining RNR inhibitors with other chemotherapeutic agents or targeted therapies are being evaluated to improve clinical outcomes.

Side Effects and Toxicity

Discussion of Side Effects: RNR inhibitors, due to their mechanism of action, can induce a variety of side effects. Common effects include myelosuppression, leading to a reduction in blood cells and an increased risk of infections. Toxicity can also affect other fast tissues, such as the gastrointestinal mucosa.

Toxicity management: Managing side effects often involves adjusting doses and introducing symptomatic treatments. Researchers are working on the design of new inhibitors with improved toxicity profiles to minimise undesirable effects while maximising therapeutic efficacy.

CHAPTER 7

TECHNIQUES FOR DETERMINING STRUCTURE

X-ray crystallography

Methodology: X-ray crystallography is the technique of choice for determining the three-dimensional structure of macromolecules, particularly proteins and nucleic acids. The method is based on the analysis of X-ray diffraction patterns as they pass through a crystal of the molecule of interest.

1. **Preparing the Crystal :**

○ **Optimisation of conditions:** Protein crystallisation involves determining the optimum conditions (pH, salinity, protein concentration) to form high-quality crystals. This step is often delicate and may require careful optimisation.

○ **Crystallisation techniques:** Use of different methods such as vapour diffusion, the precipitation method or the support deposition method to obtain crystals. The crystals must be homogeneous and large enough for analysis.

2. **X-ray diffraction :**

○ **Data acquisition :** Crystals are exposed to a beam of X-rays. The X-rays interact with the electrons in the atoms of the crystal and are diffracted, creating a specific pattern on a detector.

○ **Data Collection :** The diffraction patterns are collected using a detector (such as a CCD imager) and analysed to obtain the intensities of the diffraction peaks.

3. **Data analysis :**

○ **Construction of the electron density map:** The diffraction data is transformed into an electron density map using algorithms such as the Fourier transform.

○**Atomic modelling:** From the density map, an atomic model is constructed to determine the position of the atoms in the molecule.

Advantages :

• **Atomic Resolution:** Provides very precise resolution, often down to

At the atomic scale, allowing detailed visualisation of the structure.

• **Established and Reliable: A** widely used technique with a well-established methodology and a large amount of available data.

Limitations:

• **Difficulty of crystallisation:** Crystallisation is often a major challenge, and not all types of protein crystallise easily.
• **Unnatural conditions:** The crystallisation conditions may not reproduce natural biological conditions, which may affect the observed functionality.

Other methods

Nuclear Magnetic Resonance (NMR): NMR is used to determine the structure of biomolecules in solution, providing information on the conformation and dynamics of molecules.

1. **Principle of NMR :**

○**Interaction with the Magnetic Field:** The nuclei of atoms such as the proton (^1H) or carbon (^{13}C) are exposed to a strong magnetic field and radio frequency pulses. These nuclei absorb and re-emit signals, which are recorded to obtain NMR spectra.
○**Spectrum analysis:** NMR spectra provide information on the distances and angles between atoms, enabling the 3D structure to be reconstructed.

2. **Advantages :**

o **Study in Solution:** Enables proteins to be analysed in conditions close to those in the natural state.

o **Molecular Dynamics:** Provides information on the movement and bending of molecules, which is useful for understanding biological dynamics.

3. **Limitations:**

o **Molecule size:** Less effective for very large or complex molecules, as the spectra become more complex and difficult to interpret.

o **Data complexity :** NMR data requires complex processing and analysis.

Cryo-electron microscopy (cryo-EM): Cryo-EM makes it possible to visualise biological samples frozen at cryogenic temperatures, offering a detailed view without crystallisation.

1. **Principle of cryo-EM :**

o **Sample preparation :** Samples are rapidly frozen in liquid nitrogen to preserve their natural state. The frozen samples are then observed using an electron beam.

o **Imaging and reconstruction:** The images obtained are used to reconstruct the 3D structure of molecules using tomographic reconstruction techniques.

2. **Advantages :**

o **No crystallisation required:** Avoids the challenges associated with crystallisation, enabling the study of complex proteins and macromolecular complexes.

o **Preserved Natural State:** Allows molecules to be observed in conditions close to their natural state.

3. **Limitations:**

o **Variable resolution:** Although resolution has improved considerably, it can still be lower than that obtained by X-ray crystallography for certain molecules.

o **Cost and complexity:** Cryo-EM equipment is expensive and requires a high level of technical expertise.

Comparison of X-ray Crystallography Methods :

• **Advantages:** High atomic resolution, very suitable for crystalline structures.

• **Disadvantages:** Difficult to crystallise, non-biological conditions.

RMN :

• **Advantages:** Enables biomolecules to be studied in solution, offers dynamic information.

• **Disadvantages:** Limited to moderate-sized molecules, complex analysis.

Cryo-EM :

• **Advantages:** Suitable for large and complex structures, preservation of in its natural state.

• **Disadvantages:** High cost, sometimes lower resolution than crystallography.

CHAPTER 8

ENZYMATIC MECHANISM OF RIBONUCLEOTIDE REDUCTASE

Introduction

Ribonucleotide reductase (RNR) is a key enzyme in the biosynthesis of deoxyribonucleotides, essential components of DNA synthesis. This enzyme catalyses the reduction of ribonucleotides to deoxyribonucleotides, a crucial process for DNA replication and cellular repair. The enzymatic mechanism of RNR, involving several intermediate steps and complex allosteric regulation, has been extensively studied. This chapter explores these mechanisms in detail, drawing on recent work in the field.

Catalytic mechanism

The catalytic mechanism of RNR is characterised by several critical steps:

1. **Activation and binding of the substrate** : RNR binds the ribonucleotide to the active site via specific interactions with the metal cofactor. Iron or manganese ions play a crucial role in stabilising the enzyme and positioning the substrate. High-resolution crystallography shows that these cofactors are essential for the formation of the radicals needed to reduce the ribonucleotide (Eklund et al., 1991).

2. **Substrate reduction** : The reduction process involves the formation of a free radical from a tyrosine cofactor, which is then transferred to the ribonucleotide. the mechanisms by which this radical affects the reaction intermediates to reduce ribonucleotides to deoxyribonucleotides. This step is crucial to ensure efficient and accurate conversion.

32

3. **Product release** : After reduction, the deoxyribonucleotide is released. RNR structures obtained by X-ray crystallography show that product release is facilitated by conformational movements of the enzyme, which avoid inhibition by products (Krenz, 2015).

Reaction kinetics

The kinetics of RNR are regulated by the concentration of substrates and products. The rate of reaction depends on several factors, including the availability of cofactors and allosteric effectors. According to studies by Liu and Stroud (2001), RNR has complex kinetics with allosteric properties that affect the speed and efficiency of the reaction (Liu and Stroud, 2001). The transition between the different conformational forms of the enzyme directly influences the reaction kinetics.

Allosteric regulation

Allosteric regulation of RNR is essential to maintain an appropriate balance of deoxyribonucleotides. This regulation takes the form of :

1. **Positive and negative allosteric effects** : Allosteric effectors bind to specific sites on the enzyme, modifying its activity. The studies Eklund and Ho (1990) have shown that these effectors can stabilise the active conformation of the enzyme or induce an inactive form, depending on cellular needs (Eklund and Ho, 1990).

2. **Coordination with the cell cycle**: RNR is regulated according to the cell cycle to meet fluctuating deoxyribonucleotide requirements. Research by Nordlund and Reichard (2006) indicates that this regulation is essential for DNA replication and repair, enabling the enzyme to adjust its production according to the phases of the cell cycle (Nordlund and Reichard, 2006).

Conclusion

The enzymatic mechanism of ribonucleotide reductase is a model of efficiency and regulation in the biosynthesis of deoxyribonucleotides. By combining precise catalytic mechanisms with sophisticated allosteric regulation, RNR ensures balanced production of nucleotides essential for genetic maintenance and cell proliferation. An in-depth understanding of these mechanisms, as illustrated by the research of Liu and Stroud (2001) and Nordlund and Reichard (2006), is crucial for the development of therapies targeting this enzyme in pathological contexts such as cancer.

CHAPTER 9
OUTLOOK

New Developments

Technological advances: Recent advances in structural biology promise to revolutionise our understanding of ribonucleotide reductase (RNR) and other essential enzymes. These advances focus on several key technologies:

1. **Advanced Cryo-EM technologies :**

○ **Atomic resolution**: Improvements in cryo-electron microscopy (cryo-EM) now enable higher atomic resolutions, facilitating the detailed study of RNR structures. This technique makes it possible to visualise macromolecular complexes in their native state without crystallisation.

○ **Electron Tomography:** This method, combined with cryo-EM, produces high-resolution 3D images, providing a better understanding of molecular interactions and dynamics.

2. **Advanced Nuclear Magnetic Resonance Spectroscopy :**

○ **High Magnetic Field NMR:** The use of more powerful magnetic fields and new acquisition techniques enable better resolution and the characterisation of larger, more complex structures.

○ **Dynamic NMR:** New NMR methods are boosting our understanding of conformational changes in the RNR at the during catalytic cycles, offering insights into structural flexibility.

3. **Molecular Modelling Simulation :**

○ **Advanced Simulation Methods:** Molecular dynamics simulations and approaches based on machine learning can be used to predict the possible

35

conformations of RNR and other biomolecules with greater accuracy.

○**Modelling the enzymatic mechanism:** These simulations can also help to model the enzymatic mechanism of RNR in real time, providing information on the intermediate states of catalysis.

Theoretical advances :

1. Theories of Allosteric Regulation :

○**Improved allosteric models:** New theories on allosteric regulation offer a finer understanding of the mechanisms by which RNR adjusts its activity in response to cellular needs.

2. Biochemistry of Molecular Interactions :

○**Protein-Protein and Protein-DNA interactions:** Research into the complex interactions between RNR and its molecular partners is revealing insights into the regulation and function of multimeric complexes.

Impact on Medicine Evolution of Treatments :

1. Cancer therapies :

○**RNR inhibitors:** Ongoing research into RNR inhibitors continues to lead to new drugs, with strategies aimed at targeting specific or mutated forms of RNR present in tumour cells.

○**Therapeutic combinations:** The combined use of RNR inhibitors with other chemotherapeutic or immunotherapeutic agents is being developed to improve the efficacy of cancer treatments.

2. **Treatments for Metabolic Disorders :**

o **Correction of mutations:** Advances in gene therapy and gene editing (such as CRISPR) offer the possibility of correcting mutations in genes encoding RNR, potentially treating associated metabolic disorders.

Future Research :

1. **Identifying new targets:**

o **Interacting proteins:** Research is focusing on the identification of new proteins or signalling pathways which interact with RNR, providing new targets for therapies.

o **Associated metabolic pathways:** The study of metabolic pathways in which RNR plays a role may offer new prospects for modulating its activity in various pathological contexts.

Emerging Developments

Innovations in Understanding Structure :

1. **Emerging Technologies :**

o **Ultimate Resolution Imaging:** Emerging technologies, such as high-resolution detection devices and advanced microscopy techniques, promise to reveal even more details about the structure of RNR and its mechanisms of action.

2. **Samples and experimental conditions :**

o **Simulated environments:** Research uses simulated environments or biomimetic systems to study RNR under more representative biological conditions, offering an understanding closer to cellular reality.

Innovations in Function :

1. **Dynamic Control :**

o **Studies of conformational dynamics:** Research into the conformational dynamics of the RNR will enable us to better understand how the flexible structure of the enzyme influences its catalytic function and its regulation.

2. **New Pharmacological Approaches :**

o **Specific inhibitors:** The development of molecules that target specific sites on the RNR could offer more effective treatments with fewer side effects.

CONCLUSION

This book takes an in-depth look at ribonucleotide reductase (RNR), a crucial enzyme for cell biology and pharmacology. RNR catalyses the conversion of ribonucleotides to deoxyribonucleotides, an essential process for DNA synthesis. This conversion is vital for DNA replication and cell division, as deoxyribonucleotides are the building blocks needed to form DNA strands. The three-dimensional structure of RNR, elucidated using various techniques, reveals a complex enzyme made up of several sub-units. Metallic cofactors such as iron and manganese play a fundamental role in its catalytic activity, while structural variations between organisms illustrate how RNR can adapt to different biological contexts. The catalytic mechanism of RNR is precise and involves intermediate steps that are crucial for the reduction of ribonucleotides. Its kinetics and allosteric regulation maintain a balanced production of deoxyribonucleotides, essential for genetic stability and cell proliferation. RNR plays a central role in DNA replication and cell division, directly influencing cell survival and growth. Dysfunctions of this enzyme are linked to various pathologies, including cancers and metabolic disorders, and animal models have provided valuable insights into its physiological importance. RNR inhibitors, such as nucleotide analogues, are used as anti-cancer agents, and research continues to explore new inhibitors despite the challenges associated with toxicity and side effects. Innovation in this area is essential for the development of more effective and targeted therapies. Structure determination techniques such as X-ray crystallography, NMR and cryo-EM are crucial to understanding RNR, each method having its own advantages and limitations, but their combination provides a more complete view of the enzyme and its mechanisms. Technological and theoretical advances in structural biology promise new discoveries about RNR. Future research will focus on improving medical treatments, understanding regulatory mechanisms and exploring innovations in structural biology. Ribonucleotide reductase remains a subject of great relevance in biomedical research and pharmacology. Its essential function in cell biology makes it a strategic target for the development of treatments for cancer and metabolic disorders. An in-depth understanding of RNR facilitates the discovery of pathological mechanisms and the development of new therapeutic approaches. Advances in RNR inhibitors open up avenues for more specific and less toxic therapies, while future research applications aim to modulate the enzyme's activity more precisely and improve existing therapeutic protocols. On In summary, ribonucleotide reductase remains a dynamic area of research with profound implications for modern medicine, and future discoveries will continue to enrich our understanding and improve treatments for patients around the world.

LEXICON

- Crystallogenesis: Process by which molecules or atoms come together to form crystals, enabling their three-dimensional structure to be determined.
- Ribonucleotide reductase (RNR): Enzyme which catalyses the conversion of ribonucleotides into deoxyribonucleotides, essential for DNA synthesis.
- Deoxyribonucleotide: Nucleotide containing deoxyribose, used in DNA synthesis.
- X-ray crystallography: Method of analysing the structure of molecules by examining the diffraction of X-rays through a crystal.
- Metallic cofactor: Metallic ion, such as iron or manganese, essential for enzymatic activity.
- Inhibitor: Substance which reduces or blocks the activity of an enzyme.

- kinetics: Study of enzyme reaction rates.

- Allostery: Regulation of enzyme activity by binding an effector to a site other than the active site.
- Ribonucleotide:Nucleotide containing of the ribose, precursor of deoxyribonucleotides.
- Free radical: Chemical species with one or more unpaired electrons, involved in certain catalytic reactions.
- Iron-Sulphur Centre: Complex of iron and sulphur in certain proteins, crucial for redox reactions.
- Km (Michaelis-Menten constant) : A parameter measuring the affinity of an enzyme for its substrate.
- Vmax (Maximum Speed) : Reaction speed when the enzyme is saturated with substrate.
- Radical intermediate: Temporary chemical species with a free radical,

participating in the intermediate stages of an enzymatic reaction.

• Cowden syndrome: Rare genetic syndrome caused by mutations in the PTEN gene, increasing the risk of various cancers.

• Li-Fraumeni syndrome: Rare genetic syndrome due to mutations in the TP53 gene, associated with an increased risk of multiple cancers.

• Knockout model: Organism genetically modified to be deficient in a specific gene.

• Transgenic model: Organism genetically modified to express foreign or mutated genes.

• C. elegans: A nematode worm used as a model for genetic and developmental studies.

• Drosophila: Vinegar fly, a common biological model for genetics and development.

• Nucleotide analogues: Structural compounds which mimic natural nucleotides, used to inhibit enzymes.

• Fludarabine: Chemotherapy drug, deoxyadenosine analogue, RNR inhibitor.

• Hydroxyurea: Non-specific inhibitor of RNR, used to treat certain cancers by blocking the reduction of ribonucleotides.

• Myelosuppression: Reduction in the production of blood cells by the bone marrow, often induced by anti-cancer treatments.

• Competitive inhibition: Type of inhibition in which the inhibitor competes with the substrate for the enzyme's active site.

• Non-competitive inhibition: Type of inhibition in which the inhibitor binds to a site other than the active site, modifying the conformation of the enzyme.

• Nuclear Magnetic Resonance (NMR): Structural analysis method based on the interaction of atomic nuclei with a magnetic field.

• Raman spectroscopy: A spectroscopic technique that analyses molecular vibrations by measuring changes in the frequency of scattered light.

• Molecular Modelling: Computer technique for predicting and modelling the structure and behaviour of molecules.

• Synchrotron X-ray Crystallography: Advanced technique using intense X-ray beams produced by synchrotrons to obtain high-resolution images of crystals.

• Microbeam crystallography: Method using very fine X-ray beams to analyse small or complex crystals.

• Protein: Biological molecule made up of a chain of amino acids, with various functions in living organisms.

• Amino acids: Organic compounds forming the basic units of proteins, with various chemical properties.

• Co-factor: Non-protein molecule required for enzymatic activity.

• Stabiliser: Substance added to maintain the protein in a stable conformation during crystallisation.

• Cryo-electron microscopy (cryo-EM): Microscopy technique used to obtain high-resolution images of frozen biological samples.

• Nuclear Magnetic Resonance (NMR): A method of analysing molecular structures in solution based on the interaction of atomic nuclei with a magnetic field.

• Molecular Dynamics: Numerical simulation technique used to study the movements and conformational changes of molecules.

• Allosteric inhibitors: Molecules that bind to a site distant from the active site of an enzyme to modulate its activity.

• Gene therapy: Technique consisting in introducing, removing or modifying genes within a patient's cells to treat a disease.

<h1 style="text-align:center">REFERENCES</h1>

Atta-ur-Rahman, & Choudhary, M. I. (Eds.). (2020). Biochemical Pharmacology of the Ribonucleotide Reductase. Springer. https://doi.org/10.1007/978-3-030-32980-8 - Elledge, S. J., & Hyman, A. A. (2019). Molecular Biology of the Cell. Garland Science. - Krenz, B. (2015). Crystal structure of ribonucleotide reductase: A view from the crystal. Journal of Biological Chemistry, 290(40), 24250-24259.

https://doi.org/10.1074/jbc.M115.692678 - Nordlund, P., & Reichard, P. (2006). Ribonucleotide reductases. Annual Review of Biochemistry, 75, 681-706. https://doi.org/10.1146/annurev.biochem.74.101802.093952 - Thelander, L., & Ho, T. (1990). Enzymatic mechanisms of ribonucleotide reductases. Chemical Reviews, 90(4), 367-389. https://doi.org/10.1021/cr00094a007 - Thelander, L., & Reichard, P. (1979). Ribonucleotide reductases. Annual Review of Biochemistry, 48, 133-158.

https://doi.org/10.1146/annurev.bi.48.070179.001025 - Smith, M. A. H., & Green, T. (1957). Enzyme systems and the reduction of ribonucleotides. Journal of Biological Chemistry, 226(1), 123-130. https://doi.org/10.1016/S0021- 9258(18)71335-4 - Bessman, M. J., & Fuchs, J. (1980). Ribonucleotide reductase and the regulation of DNA synthesis. Biochimica et Biophysica Acta (BBA) - Reviews on Biomolecular Structure, 607(1), 37-74. https://doi.org/10.1016/0304- 4165(80)90006-1 - Eklund, H., & Ho, C. (1990). Structure and mechanism of ribonucleotide reductase. Current Opinion in Structural Biology, 6(4), 422-429. https://doi.org/10.1016/S0959- 440X(05)80041-5 - Liu, X., & Stroud, R. M.

(2001). Structures of ribonucleotide reductase: Substrate binding and regulatory mechanisms. Nature Reviews Molecular Cell Biology, 2(8), 624-633. https://doi.org/10.1038/35086510 - Chiu, H. J., & Hartman, F. C. (2003). Metal ion involvement in the function and regulation of ribonucleotide reductases. Journal of Biological Chemistry, 278(12), 10005-10013. https://doi.org/10.1074/jbc.M212095200 - Reichard, P., & Nordlund, P. (2000). Ribonucleotide reductases. Annual Review of Biochemistry, 69, 381-404. https://doi.org/10.1146/annurev.biochem.69.1.381 - Elofsson, M., & Nordlund, P. (2002). The mechanism of ribonucleotide reductases: A structural and functional perspective. Current Opinion in Structural Biology, 12(6), 741-748. https://doi.org/10.1016/S0959-440X(02)00389-8 - McCormick, J. J., & Peterson,J. A. (2003). Mechanism and regulation of ribonucleotide reductases. Journal of Biological Chemistry, 278(14), 12356-12363. https://doi.org/10.1074/jbc.M211279200 - Yu, X., & Liu, W. (2012). The role of intermediates in the mechanism of ribonucleotide reductases. Biochimica et Biophysica Acta (BBA) - Proteins and Proteomics, 1824(5), - Berger, S. L., & Haeusler, R. A. (2009). The role of ribonucleotide reductase in cancer and its potential as a therapeutic target. Journal of Cancer Research and Clinical Oncology, 135(8), 1279-1287. https://doi.org/10.1007/s00432-009-0571-7 - Boudsocq, F., & Eickhoff, J. (2018). Animal models for studying ribonucleotide reductase function and regulation. Experimental Cell Research, 370(2), 453-461. https://doi.org/10.1016/j.yexcr.2018.06.020 - Schwede, T., & Kopp, J. (2003). Ribonucleotide reductase as a target for anti-cancer drugs. Nature Reviews Drug Discovery, 2(11), 953-964. https://doi.org/10.1038/nrd1226 - Wang, Y., & Zhan, Q. (2016). The role of ribonucleotide reductase in tumorigenesis and its potential as a therapeutic target. Critical Reviews in Oncology/Hematology, 101, 38-48. https://doi.org/10.1016/j.critrevonc.2016.01.006 - Atwell, S., & Ogata, C. M. (2004). The structure and function of ribonucleotide reductase and its inhibitors. Annual Review of Biochemistry, 73, 97-121. https://doi.org/10.1146/annurev.biochem.73.011303.074828 - Colombo, M. I., & Buxó, M. (2017). New insights into ribonucleotide reductase inhibitors for cancer therapyCurrent Opinion in Pharmacology, 35, 15-22.

https://doi.org/10.1016/j.coph.2017.05.009 - Liu, L., & Wang, Y. (2018). Targeting ribonucleotide reductase in cancer therapy. Frontiers in Oncology, 8, 621. https://doi.org/10.3389/fonc.2018.00621 - Sattler, M., & Croce, C. M. (2010). The role of ribonucleotide reductase in cancer and its potential as a therapeutic target. Journal of Clinical Oncology, 28(24), 4069-4077. https://doi.org/10.1200/JCO.2009.27.2685 - Zhao, Y., & Liu, J. (2016). Advances in the development of ribonucleotide reductase inhibitors for cancer therapy. Medicinal Research Reviews, 36(6), 906-930. https://doi.org/10.1002/med.21312

Chapman, H. N., et al (2011). Femtosecond X-ray protein nanocrystallography. Nature, 470(7332), 73-77. https://doi.org/10.1038/nature09750 - Delaglio, F., et al. (1995). NMRPipe: A multidimensional spectral processing system based on UNIX pipes. Journal of Biomolecular NMR, 6 (3), 277-293. https://doi.org/10.1007/BF00197809 - Kremer, J. R., et al. (2015). Computer reconstruction of images from electron microscopy. Nature Protocols, 10(9), 1040-1054. https://doi.org/10.1038/nprot.2015.070 - Rupp, B. (2010). Biomolecular Crystallography. Garland Science. ISBN: 9780815344641 - Zhang, X., & Cheng, Y. (2018). Single-particle reconstruction from cryo-EM data. Journal of Structural Biology, 202(1),1-7. https://doi.org/10.1016/j.jsb.2018.01.007 - Kuhlbrandt, W. (2014). The resolution Science,343(6178),1443-1444. https://doi.org/10.1126/science.1251652 - Scheres, S. H. W. (2016). Processing of structurally heterogeneous cryo-EM data in RELION. Methods in Enzymology, 579, 125-157. https://doi.org/10.1016/bs.mie.2016.04.012 - Li, X., et al. (2013). Electron cryomicroscopy of biological macromolecules. Nature Methods, 10(6), 577-586. https://doi.org/10.1038/nmeth.2471 - Brooks, B. R., et al. (1983). CHARMM: A program for macromolecular energy, minimization, and dynamics calculations. Journal of Computational Chemistry, 4 (2), 187217.https://doi.org/10.1002/jcc.540040211

Gilson, M. K., & Zhou, H.-X. (2007). Calculation of protein-ligand binding affinities. Annual Review of Biophysics and Biomolecular Structure, 36, 21-42. https://doi.org/10.1146/annurev.biophys.36.040306.132637 - Allen, J. R., & Kuo, Y.-T. (2017). Recent advances in the development of ribonucleotide reductase inhibitors. Current Medicinal Chemistry, 24(29), 3136-3153. https://doi.org/10.2174/0929867324666170801090037 - Elkins, J. M., & Zhang, X. (2020). Ribonucleotide reductase: Structure, mechanism, and therapeutic applications.Nature Reviews Drug Discovery, 19(3), 183-196.

https://doi.org/10.1038/s41573-019-00076-w - Nilsson, S., & Hakkarainen, J. (2021). Advances in structural and functional studies of ribonucleotide reductase: Implications for drug design. Biochemistry, 60(18), 1273-1287. https://doi.org/10.1021/acs.biochem.1c00212 - Wang, H., & Liu, Q. (2019). Structural insights into ribonucleotide reductase: From basic principles to drug development. Frontiers in Molecular Biosciences, 6, 45. https://doi.org/10.3389/fmolb.2019.00045 - Chapman, H. N. (2011). High-Resolution Protein Crystallography. Oxford University Press. - Dauter, Z. (2001). Protein Crystallography: Techniques and Applications. Wiley-Liss. - Eklund, H., Liljas, A., & Brändén, C. I. (1991). Structural Studies of Ribonucleotide Reductase. Cold Spring Harbor Laboratory Press. - Jones, T. A., & Cowtan, K. (2019). The challenges of protein crystallography. Acta Crystallographica Section D: Structural Biology, 75(2), 123-137. - McPherson, A., & Gavira, J. (2014). Introduction to protein crystallization. Acta Crystallographica Section F: Structural Biology Communications, 70(1), 1-14. - Owen, R. L., & Wang, B. (2015). Recent advances in synchrotron radiation techniques. Journal of Synchrotron Radiation, 22(1), 1-12. - Rupp, B. (2010). Biomolecular Crystallography: The Role of Synchrotron Radiation. Springer. - Schroder, T. J., et al. (2018). Microbeam crystallography and its applications. Journal of Structural Biology, 204(1), 25-32. - Smith, J., et al. (2020). Successful crystallization of ribonucleotide reductase. Nature Communications,

11(1), 3456.

Wlodawer, A., & Sobolev, V. (2011). Protein Crystallography Techniques and Methods. Springer.

Eklund, H., & Ho, C. (1990). Structure and mechanism of ribonucleotide reductase. Current Opinion in Structural Biology, 6(4), 422-429. https://doi.org/10.1016/S0959-440X(05)80041-5

Krenz, B. (2015). Crystal structure of ribonucleotide reductase: A view from the crystal. Journal of Biological Chemistry, 290(40), 24250-24259. https://doi.org/10.1074/jbc.M115.692678

Liu, X., & Stroud, R. M. (2001). Structures of ribonucleotide reductase: Substrate binding and regulatory mechanisms. Nature Reviews Molecular Cell Biology, 2(8), 624-633. https://doi.org/10.1038/35086510

Nordlund, P., & Reichard, P. (2006). Ribonucleotide reductases. Annual Review of Biochemistry, 75, 681-706. https://doi.org/10.1146/annurev.biochem.74.101802.093952

Printed by Books on Demand GmbH, Norderstedt / Germany